SCI-FI EXPLAINED

MIND AND MATTER

Michael White is the author of many science books for adults and children, including the bestselling *Stephen Hawking – A Life in Science* (with John Gribbin). He has written books about Albert Einstein, Charles Darwin and Isaac Newton.

Michael is also a big fan of science fiction. Some of his favourites are *The X-Files* and *Star Trek* and the films *2001: A Space Odyssey* and *Bladerunner*.

Another book in this series

SCI-FI EXPLAINED:
ALIEN LIFE-FORMS

Sci-Fi Explained

Mind and Matter

Michael White

PUFFIN BOOKS

PUFFIN BOOKS

Published by the Penguin Group
Penguin Books Ltd, 27 Wrights Lane, London W8 5TZ, England
Penguin Putnam Inc., 375 Hudson Street, New York, New York 10014, USA
Penguin Books Australia Ltd, Ringwood, Victoria, Australia
Penguin Books Canada Ltd, 10 Alcorn Avenue, Toronto, Ontario, Canada M4V 3B2
Penguin Books (NZ) Ltd, Private Bag 102902, NSMC Auckland, New Zealand

On the World Wide Web at: www.penguin.com

Penguin Books Ltd, Registered Offices: Harmondsworth, Middlesex, England

First published 1999
1 3 5 7 9 10 8 6 4 2

Set in Helvetica

Made and printed in England by Clays Ltd, St Ives plc

British Library Cataloguing in Publication Data
A CIP catalogue record for this book is available from the British Library

ISBN 0–141–30017–5

Contents

Introduction

Are you a sci-fi fan? Do you enjoy watching films and reading books about subjects like future worlds, time travel and superhuman powers?

Science fiction is great entertainment. Sci-fi writers are able to use their imaginations and really let loose with their ideas about people and places. This is something that *real* scientists can't do. Real scientists have to work with the evidence that is there in front of them, and they can't make things up. Real science works with sets of rules, and these rules cannot be broken.

But many scientists are big sci-fi fans, too. In fact, some scientists say that sci-fi is the reason they got interested in science when they were younger. The famous physicist Professor Stephen Hawking even appeared in an episode of one of the longest-running sci-fi series ever, *Star Trek: The Next Generation*!

But the difference between the imaginary worlds of sci-fi and the real world of science isn't always as great as you may think. Quite often, science-fiction writers have guessed what is going to happen in the future before the real scientists have gathered the evidence. Inventions and ideas that have been dreamt up in the imagination of a writer have eventually become reality.

In *Mind and Matter* we'll look at some of the incredible superhuman powers and abilities that you'll recognize from stories in sci-fi books, films and TV programmes – and we'll find out what the real scientists have to say.

There are lots of questions which we don't know the answers to yet. But one thing is certain: no matter how weird science fiction may be, the reality can sometimes be just as strange . . .

Chapter 1

Ghost Hunting

That evening the pair set up their equipment and wait.
And wait . . .
4
At 2.30 a.m. a sudden coldness fills the room and the agents wake up.
Look at that!
Quick, get this on tape!

The nun walks across the room and disappears, leaving the two agents speechless – and without any evidence.
In the morning the two tired agents pack up and leave.
I wonder what really happened at that house? Were we just hallucinating?
After last night, are you telling me you still don't believe in ghosts?
I'll have to get back to you on that one . . .
5

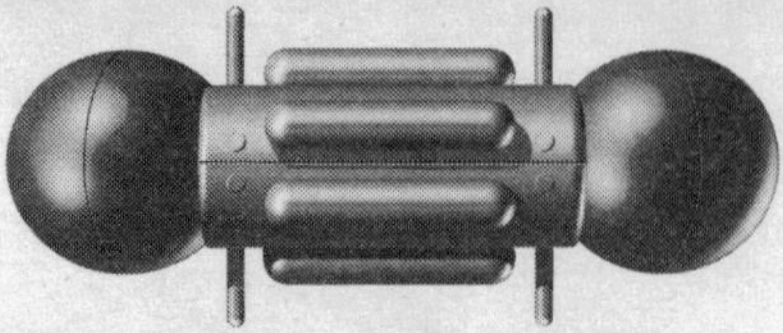

Ghost Hunting

THE FACTS

If you've ever stayed the night in a spooky house or walked through a graveyard in the dark, the thought of seeing a ghost must have crept into your mind. The idea of seeing a real ghost is both terrifying and thrilling – and most of us love telling spooky stories and watching TV programmes that scare us silly.

But if we leave all the stories and films behind, there is a real world in which many people genuinely believe that they have seen or contacted a spirit – and in which paranormal investigators are busy researching and trying to find explanations for these strange events. According to an investigator, one person in six believes in ghosts, and one person in fourteen claims to have actually seen a ghost! But can they really exist outside the world of science fiction and horror?

A Brief History of Ghosts

People have believed in ghosts from the earliest times. It is thought that cavemen put their trust in the spirits of their forefathers to help them hunt. In their world, everything had its own spirit, from the trees and plants to the animals and birds. The ancient Egyptians had a special book, called *The Book of the Dead*, which guided people's spirits on their journey in the afterlife. Some of the very earliest ghost stories come from these times. The literature of the Greeks and Romans is full of stories about ghosts, and this tradition has continued into more modern writing. One of the most famous stories ever written – the play *Hamlet* by William Shakespeare – features a ghost, Hamlet's father.

In Britain, the whole idea of ghosts really took off in Victorian times, when it became very fashionable to go on 'ghost hunts' and to write about strange experiences in what were believed to be haunted houses.

Towards the end of the last century, a group of academics got together and formed what they called the Society for Psychical Research (shortened to the SPR). The aim of the group was to find a scientific explanation for the existence of ghosts. At about this time, it was considered the height of fashion to be seen attending seances. A seance is a gathering in which a person called a medium (who claims to have special psychic powers) tries to contact the dead in order to allow a 'conversation' to take place between them and their living friends and relatives.

The SPR believed that such seances were nothing but a trick, and they did their best to expose fakery. Some mediums were shown to be tricksters, and a few were even arrested and imprisoned. The fashion for seances eventually died out, but even today some people still attend them in the hope of contacting relatives or friends who are no longer alive. The SPR were the earliest known example of what we now call 'paranormal' investigators. These days, other kinds of 'ghostbusting' investigators are on the scene.

Today's Paranormal Investigators

The SPR still exists today and has over a thousand members; they attend lectures and workshops covering a range of paranormal activities, including, of course, studying the subject of ghosts. There is also the ASSAP (short for the Association for the Scientific Study of Anomalous Phenomena), a smaller,

Lots of photographs have been taken of 'ghosts' over the years, though many have proved to be fakes. However, people remain baffled by this famous shot, taken at Raynham Hall, Norfolk in 1936.

more scientific group that carries on similar investigations. As well as this, there are university departments around the country in which scientists called parapsychologists carry out research into all kinds of phenomena, including ghost and poltergeist activity. (See Chapter 2, *The Power of Thought*, for more on the Koestler Institute.)

But investigation of the paranormal is not just for the believers. There are many groups of disbelievers (called sceptics) who try to use scientific methods to find out what the phenomenon called 'a ghost' could really be. The most important of these is the Committee for the Scientific Investigation of Claims of the Paranormal. One of its founder members is a magician called James Randi; he has spent a lot of of time trying to show that many paranormal stories and reports are untrue. He believes that every 'supernatural' event he has ever witnessed can be explained by trickery – the kinds of tricks and illusions he himself uses in his stage shows. However, many people think that Randi's 'explanations' leave many paranormal mysteries unsolved.

What is a Ghost?

Many people believe that ghosts are simply the

spirits of the dead: when someone dies, they may feel that they've left behind 'unfinished business', so they hang around in a 'ghostly' form to try to put things right – which is why they are often described as 'restless souls'. However, there is no scientific evidence for this and, as far as we know, no scientist has ever managed to hold a conversation with a ghost to find out exactly what it is they're up to!

But if ghosts are not images of long-dead people, what could they be? For many years investigators of the paranormal have tried to pin a ghost down and find out what is special about the places where ghosts are seen. As you'll find out, there is no shortage of different theories to try to explain their existence.

Optical Illusions

Have you ever looked at an object and, for a few seconds, thought it was something completely different? Things can often appear to be something they're not: a coat hanging up can suddenly become a human figure, a gnarled branch can turn into an arm. This trick played by the eye is known as an optical illusion. This is one possible explanation for people seeing 'ghosts'.

Other kinds of optical illusions can be created by artists in pictures that deliberately fool the brain into seeing something other than what is really there. Another type of illusion is a mirage, a trick of the light usually seen in hot, dry places like deserts. The light from a distant object can be bent or the image become twisted, so that people think they are seeing things that aren't there. Some travellers claim to have seen an oasis in the desert when there was nothing there but dry sand – an effect which may also be heightened when the traveller is exhausted and thirsty. In this state, your mind as well as the light can play tricks on you.

Many people claim to have seen ghosts at night or in half-lit conditions. A common setting for a ghost is a dark place like a graveyard or an old building shrouded in mist. Such sightings are often shown to be illusions because these dark and dingy conditions can easily convince people that they have seen something spooky. People are also more likely to believe they are seeing something strange because of the creepy atmosphere. It is easy to see how you could believe you had seen a ghost if you were wandering around a creaky old house filled with strange shadows.

Hallucinations

Hallucinations are a bit different from optical illusions. These are visions created in the mind by people who really want to believe or who are so scared that they frighten themselves into believing they can see something. Often, if two or three people are together in a spooky place, one of them will put an idea into the others' heads, perhaps by telling a ghost story. This then scares everyone and, quite unintentionally, they start to exaggerate what they are experiencing. Sometimes a group of people will scare one another so much that an innocent event can escalate into something weird or extraordinary. This is called a 'hysterical response' (see *Poltergeist in the House*, page 25).

Spirits in the Night

One of the most common times to see a ghost is when you are going to sleep or if you wake up suddenly during the night. Scientists have a very good explanation for this.

Your body's responses are dictated by your brain, which controls all the nerves throughout your body. When you go to sleep, this nervous system steps down. While asleep you don't need many of the systems that operate automatically during your waking hours: you don't need to be alert or to be so aware of your surroundings, you don't need to move or touch or smell or see. All you need is a basic operating system that allows you to breathe, digest your last meal and perform other automatic processes (as well as letting you be sufficiently aware of your surroundings to wake up in case of a serious danger, like a fire). Because of this we actually have two types of nervous system that control our bodies. One is called the 'voluntary nervous system'; this operates during our waking hours. The other is the 'involuntary nervous system'; this operates when we are asleep.

Have you ever had a strange vision or experience just as you were about to go to sleep at night? One of the most common is to feel as though you are about to walk off the top of a cliff or fall down a deep well. As you experience this, you give a jolt in your sleep and suddenly wake up. Your nervous system is responsible for this. Sometimes your brain gets confused when you switch from the voluntary nervous system that you've been using all day to the involuntary system – the 'night watchman'. This results in a spasm, a muscular movement that you can't control. One part of your brain tries to 'explain' this to another part (your reasoning centre) by creating an imaginary scene such as falling off a cliff!

Scientists now believe that this 'confusion' in the brain can also produce images from deep down in our minds or our memory – which would explain many supposed ghost sightings. In many cases, ghosts appear at the foot of the bed or by the moonlit window in a strange bedroom. Many of these experiences are types of hallucinations which occur either as you drop off to sleep or when you wake up.

Ghostly Hitchhikers

Hallucination could also explain some of the experiences that drivers have had over the years which involve 'ghostly' hitchhikers. There have been quite a few reports of drivers seeing ghostly people standing at the side of the road, particularly on stretches of road where a serious accident once happened. Some drivers have even reported giving a lift to a person – only to find that the hitchhiker then mysteriously disappears from the back seat of their car! Are these really the ghosts of people killed in car accidents?

Scientists can actually put many of these cases down to drivers falling asleep at the wheel and then having a hallucination as they wake up, in the same way as you might do in bed. Driving along an eerily dark road on your own would also put you in the right frame of mind to experience something spooky.

Playbacks

We've looked at some rational explanations for sightings of ghosts. But not all of them can be explained in these ways. Some 'ghosts' defy all explanation: they are not seen in spooky graveyards, they are not witnessed by hysterical or scared people, they are not visions that appear as you go to sleep or wake up. Is there any scientific explanation for such events?

Let's look at one particular case, a famous incident which took place at a girls' school in Cheltenham during the 1930s.

The New Year's Eve Nun

On this particular New Year's Eve, the headmaster of the school and a nurse both saw what appeared to be the ghost of a nun at the edge of the school playground. Curiously, the nun was in a sitting position, but there was no chair under her. A year later, on the following New Year's Eve, the pair again saw the very same nun sitting on her 'invisible' chair.

Some people have said that it was more than an interesting coincidence that both sightings happened on New Year's Eve – when there may well have been a few drinks available in the staff-room; but there may be another explanation.

The first thing to notice is that the nun paid absolutely no attention to the 'sighters' – the people who had seen her. She was sitting as if she was in her own home, and she did not seem to be able to see any of the people in the playground. It was as if she had been captured in time and her image 'played back' in the place where her room might once have stood. In fact, the nun had appeared in the form that ghost-researchers refer to as a 'playback image'. This is quite a common type of ghost. Various witnesses often report seeing the same ghost appearing in the same place. The ghost is always repeating the same action and seems completely unaware of its new surroundings. In some cases, ghosts are seen to walk through doors that no longer exist or upon floors that have long since been demolished.

In the Treasurer's House in York over the years there have been many sightings of a group of tired-looking Roman soldiers marching through the cellar and disappearing into the walls. One theory is that the house may have been built on the site of a former Roman road which the soldiers had been walking along in their own time.

The basic idea behind the 'playback' image is that somehow an event can become 'imprinted' on its surroundings. In the case of the nun, when she was sitting in that particular chair in her own time, for some odd reason her image was 'preserved' in the air or in the walls of the building. Then, at a future date, the image gets played back, rather like a video. It is difficult to work out exactly how this could happen, but there are some possible explanations.

Recordings from the Past?

All forms of recording, CDs and videos for example, need to be imprinted on what is called a 'medium', such as a cassette, a compact disc or a video tape. (People who lead seances and try to get in touch with ghosts are also called 'mediums', because they act as a means of communication.) A photograph is produced when light 'switches on' special chemicals on photographic paper (the medium), and the photo can then be reproduced as many times as you like. A CD recording of your favourite band is made by creating a series of patterns on a plastic disc. When the laser beam in the CD-player scans these patterns, the original sound is reproduced.

If we try to explain the appearance of ghosts in a similar way, we hit a few snags. First, no machine is involved. To watch a video you need a video player; to listen to your favourite CD you need a stereo system. In the case of played-back ghost images, there is no recording device and no playback machine. One theory that ghost investigators are trying to find out more about is that the material used to construct old buildings is somehow able to store the 'information' from a past scene and then release it again at a future date when the conditions are right.

The second problem is that the preserved energy of a ghost seems to be linked with human emotions in some way. Scientists know of no way in which 'emotional energy' from humans can be stored and played back.

Finally, we don't know how ghosts receive the energy needed to power such an event. All machines, such as video-players and CDs, need an energy source, usually electricity.

Ghosts are generally seen moving, so there must be something in the way their images are stored that allows us to see them moving. One possible explanation is that the 'sighter' – the person seeing the ghost – provides the energy. Another idea is that energy is somehow taken by the ghost from its surroundings.

Although no one really knows exactly how playback images could produce sightings of ghosts, there are some interesting aspects of ghosts which seem to fit in with this idea:

- **People who see ghosts often report a sudden drop in temperature and a feeling of coldness immediately before they see the apparition. This fits in with the idea that the ghost 'draws' the energy it needs from its surroundings.**

- **Ghost images seem to fade with the passing of time. It is much more common to encounter ghosts from Victorian times than from more ancient periods, such as Roman or medieval times. This is just what you would expect from a 'recording': videotapes wear out and records get scratched.**

- **Some people, such as mediums, claim they are psychic and are in touch with the**

spirit world. Not all of them are necessarily fakes. If some people are more sensitive than others to such things, it may be possible that they have some kind of ability to 'unlock' the stored image of a ghost.

There is also a theory that ghost images or playbacks are produced only when some dramatic event has occurred; something so powerful that a form of 'emotional energy' is produced which lasts for many years. This could explain why ghosts are often images of people who are very upset about something. But if this is the case, shouldn't we expect to find more ghosts in places where terrible things have happened in the past?

In some cases this seems to have happened. In Hong Kong, the greatest number of ghosts are reported in buildings that were occupied during the Second World War, and in particular those in which Chinese prisoners were killed. The second highest population of ghosts in the city is to be found in the hospitals. But if human beings were able to 'imprint' their feelings on to the environment, why are the places where the first atomic bombs were dropped (the Japanese cities of Hiroshima and Nagasaki) not haunted by powerful after-images of the thousands who died there? Why are the sites of the Second World War concentration camps in Europe not haunted by millions of ghosts? Could it be that the environmental conditions in those places were not right at the time to imprint images, or that the conditions to 'play' them back are not suitable today?

Crisis Apparitions

Some investigators of the paranormal believe that in a time of great crisis (especially at the moment of death) some people can 'project' an image of themselves into the minds of people they know. When this happens, the person in trouble appears in the form of a ghost.

Hinchliffe's Ghost

One of the most famous cases of crisis apparition took place in the 1930s. One stormy, freezing night, a pilot called Hinchliffe and a female colleague of his were attempting to cross the Atlantic Ocean in a tiny two-seater biplane. They hit a terrible storm, the compass was damaged and they became completely lost. They radioed for help but were caught up in high winds. The plane was tossed around and began to nose-dive towards the sea. A few moments later, it hit the water and they were both killed.

The same night, two friends of Hinchliffe, Squadron Leader Rivers Oldmeadow and Colonel Henderson, were steaming towards New York on board an ocean liner. They were several hundred miles away from the scene of the crash. Neither of them had seen Hinchliffe for months, and nor did they know that he and his female companion were attempting to fly across the Atlantic.

It was in the middle of the night, at exactly the moment when Hinchliffe's plane hit the storm, that Colonel Henderson suddenly woke up and ran into his friend's room shouting: 'God, Rivers, something ghastly has happened. Hinch has just been in my cabin … It was ghastly. He kept repeating over and over again, "Hendy, what am I going to do? What am I going to do? I've got the woman with me and I'm lost. I'm lost." Then he disappeared in front of my eyes. Just disappeared.'

It seems incredible to think that Hinchliffe could somehow have projected his 'ghost' hundreds of miles away in his friend's cabin. This incident seems like a good example of a crisis apparition, but there are also some scientific explanations to consider.

First, Colonel Henderson could have been hallucinating. After a pleasant evening at the captain's table and a little too much brandy, this is quite possible. But, argue the investigators of the paranormal, how does this explain the fact that it happened on the very night that Hinchliffe died – when Henderson had no idea his friend was in the middle of a risky flight?

The answer is that it's possible that he *did* know about it, 'subconsciously' (without realizing it). If Henderson had been reading a newspaper, even weeks earlier, it is very likely that it contained an article about his friend's planned journey. Even though he didn't actually read the story, his brain could have made a note of it without it registering in his 'conscious' mind. This could have made his hallucination even more real.

So, are crisis apparitions just a bizarre coincidence, to be explained away by the fact that your brain 'knows' more than you think you know? Scientists would probably agree that this is so – but many people believe that these strange stories can't be put down to just coincidence. A question mark remains over these apparitions.

Poltergeist in the House

A poltergeist is believed to be a special type of ghost – one which behaves very badly and causes havoc in people's homes. There are quite a few reported cases of people having their homes disturbed by 'unhappy spirits'. As well as this, there have been many Hollywood films about poltergeists, mainly because the idea makes for very exciting visual scenes which are perfect for scaring cinema audiences. But what lies behind the idea of a poltergeist? Are they for real?

To the people who have been troubled by a poltergeist, they seem very real indeed. Many cases occur in family homes, and it has been suggested that the whole thing is caused by a group 'hysterical response' (see *Hallucinations*, page 13). When people get very scared, they can unintentionally trick themselves into seeing all sorts of weird things in the 'haunted house'. Any little thing that happens is exaggerated through fear and anxiety – and is blamed on an unfriendly ghost. So, the scientific explanation is that poltergeists are really produced in the minds of the people being tormented.

Another interesting idea comes from the fact that poltergeists often appear in houses where young teenagers live – especially girls. There is a theory that, when young people are growing up, which can be a time of very strong feelings and emotions, their brains produce a special form of energy. In some cases, this energy can actually take on a material form, in other words, produce very *real* effects. In a strange kind of way, the teenager is thought to be the cause of all the trouble in the house. Usually this takes the form of pipes bursting, doors slamming and furniture flying across the room – and if this is caused by the teenager, he or she doesn't do it deliberately. Scientists have no explanation for this theory, as the idea of the brain producing such large amounts of energy like this doesn't fit in at all with the laws of physics.

Another theory is that, in extreme cases, the huge amount of stress caused by a troublesome teenager in the house makes everyone in the family start hallucinating. It is hard to believe that a group of people could imagine such things, but the brain is an amazingly complex organ – and we do not yet understand all its capabilities.

Despite much investigation into this area, one thing has been discovered. If poltergeists do exist, they don't like performing in front of investigators. In many cases, the strange activities stop once the parapsychologists turn up, making it virtually impossible for them to get any real evidence on camera.

Voices From the Dead?

Back in 1920, a very famous scientist of the day – the man who invented, among other things, the light bulb and the record player – claimed he was working on a device that could be used to communicate with the dead. The man was Thomas Edison. When he announced his intention, most of his colleagues thought he had finally gone mad – but Edison was convinced he could do it. Unfortunately, he died before he could complete his project.

Three decades later, in 1959, a Swedish opera singer named Friedrich Jurgenson was listening to tape recordings of birdsong when he thought he heard human voices speaking in Norwegian in the background. Since then, many people have claimed to have heard ghostly voices speaking in the noise from a radio that has been badly tuned, or in recordings made from microphones left in apparently empty rooms. Could this be Thomas Edison's original idea brought to life?

It is difficult for scientists to accept that these recordings carry messages from the spirit world, and there are some very good reasons why. You and I are able to interpret what we see and hear in any way we like. One person can think they have heard something that another person would hear as being completely different – though they would *both* be listening to exactly the same recording. And, just as we discovered with hallucinations, one person can 'suggest' that a sound is there, and others will follow the lead; suddenly they will all be able to hear the same sound too.

Another explanation for these voices is the fact that all human beings seem to have a natural ability to find order or some kind of pattern, even in the most confusing circumstances. So if you were listening to a lot of crackle on the radio, your brain would automatically try to make sense of the noise and translate it as speech. In the same way, when we look at ink blots, which are just a random series of shapes and squiggles, we have a tendency to find pictures of things we can recognize, such as faces – again, your brain trying to make order out of a mess. This is thought to be a deep-rooted and ancient survival tool – a way in which we humans can try to make sense of things by seeking simple, understandable patterns.

'Everyday' Ghosts?

If you think about it, in a way you and I actually see 'ghosts' every day. Confused? Well, when you watch an old film, such as one starring Charlie Chaplin or Laurel and Hardy, it is almost certain that all the adult actors in it will now be dead, yet we can see them moving and talking in front of us. Another example is when we watch interviews with famous dead people, like Elvis Presley or Diana, Princess of Wales. It's a strange sensation to watch them talking about their plans and ambitions – a person who is dead, yet somehow still 'alive' on film.

Imagine how someone from the past who had never seen a film or video would feel if they were shown a recording of someone they knew who had recently died. They would certainly be forgiven for thinking they were seeing a ghost!

All in the Mind?

Quite a few ghost sightings are the results of trickery – but there are also a lot of people who genuinely believe they have seen a real ghost. Many, perhaps even all, of these cases can be put down to powerful illusions or hallucinations. One of the most interesting things to come out of all this is how little we really know and can recognize about the human mind. Its capabilities, the tricks it can play on you, and some of the things it can suggest to rational human beings can be quite incredible. So until the paranormal investigators come up with some hard evidence to prove otherwise, most scientists will assume that ghosts *are* real – but only in our own minds.

Chapter 2

The Power of Thought

Nicola concentrated on the cards.
Then harder . . .
Finally . . .
Behind the thick glass.
No way! She's a fake.
I don't think so, Jim.

I'm no fake!
Suddenly, the table went flying, though Nicola wasn't touching it.
And smashed into the glass.
Now do you believe her?

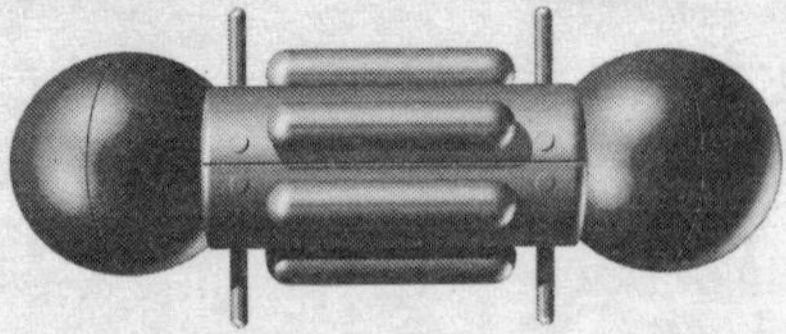

The Power of Thought

THE FACTS

While we sit on our sofas watching Mulder and Scully investigating the strange world of the paranormal, there are paranormal investigators in the real world trying to find out the truth about the power of the mind. Over the last century, hundreds of experiments have been carried out on volunteers. One team of parapsychologists in Scotland are investigating this area at the moment – and they believe their experiments show that human beings definitely *do* possess some kind of 'paranormal' powers. But many other scientists scoff at the whole idea. What exactly are they arguing about, and what kind of work is being done?

What is Telepathy?

The image from science fiction is of a person (or an alien) who has abilities well beyond those of the normal human mind: reading minds, 'seeing' the future and being able to move objects or even cause damage to them just by staring at them. In real life, some people claim that they are capable of such amazing feats. Could this be possible?

Humans normally possess five senses: sight, hearing, touch, taste and smell. There are some people, however, who claim to have something more: an extra-special 'sixth sense'. They claim that they can do things like tune into and 'read' other people's thoughts and feelings, see into the future and even communicate without speaking. The name for this special power is telepathy or ESP (short for Extrasensory Perception) and people who have this ability are called telepaths. Some telepaths even say they can 'influence objects', causing them to move or bend – and there is one person who claims he can bend solid metal objects using the power of his mind! This power is called PK (short for Psychokinesis).

For centuries people have been intrigued by the power of the mind – and the roots of telepathy may go back further in time than we think.

Telepathy – the Ancient Art?

We communicate with other human beings in lots of ways, though probably the most obvious way is by speaking. Spoken language was one of the earliest human creations and was probably first developed about 75,000 years ago. Now, of course, there are hundreds of different languages spoken in the world. But if you try to imagine yourself back in the times when we had no language, how would you have communicated with your friends and family? Some people believe that in the days before humans started using words, our ancestors used a primitive form of telepathy instead. This could have been a very advanced version of what today we call 'body language'.

Body Talk

Body language is the mass of signals we give off all the time, in our facial expressions, in the gestures of our hands and the way we stand and sit and move. It is a silent language that all of us use without realizing. An obvious example: you see someone sitting, hunched over, biting their nails, and you guess immediately, this person is nervous or worried.

As well as using body language to communicate, ancient people may also have had extremely strong senses of smell and touch to help them, far more advanced than ours are today. This meant that they may have been able to communicate effectively without the need for words. This kind of 'telepathy' could have worked well for them over short distances, but it would eventually have been replaced by a spoken language, which has many advantages. Using words, our ancestors could shout and make loud noises – they could be heard from a long way off and give warnings of danger or of predators approaching. This could not be done with body language, touch or smell – and so, according to this theory, we eventually lost our special skills.

Some parapsychologists believe that the telepaths of today may have a special link with this period in our past. A part of their brain may still have the group of cells or the special 'organ' that our ancestors used for telepathy. No one has yet discovered this, but it may be that such a feature will be found one day. The brain is an incredibly complex machine that has an amazing number of skills. It is far more advanced than the most hi-tech computer.

You and Your Senses

Most of us are fortunate enough to have the five basic senses, but some people have been able to develop one or two of them to an amazingly high level. An expert wine-taster has a very advanced sense of taste and smell, blind people gain a highly developed sense of touch when they learn to read Braille, and musicians can be trained to sing a note with perfect pitch. Each of these people has trained and developed one of his or her senses to a much higher level than an average person. Imagine what the results would be if we could train our *minds* to this level? Some people would argue that, just as Olympic athletes spend years training their bodies to be in peak condition, we could also 'train' our minds to reach their full potential. At the moment, we humans only ever use a tiny 10 per cent of our brain's capability, so there's a huge amount to be harnessed.

We are all able to communicate subconsciously (without realizing it) with each other. We can read other people's body language through their head movements and the position of their body. We can also be strongly influenced by other body 'signals', such as the tone of their voice and their smell.

If people could develop their existing senses and 'train' their minds to detect these things far better, they would then be capable of a form of telepathy called USP (short for Ultrasensory Perception).

USP and You

You may not realize it, but you are almost certainly using your own skills of USP in your everyday life. Here are two examples you might recognize.

The 'cocktail party syndrome'

Isn't it strange that, even if you're in a very loud place, such as a party or a disco, you can usually hear your name through all the noise – even if it is whispered on the far side of the room? This is an example of what is called a survival mechanism – and it is probably left over from a much earlier stage in human development. We don't think our early ancestors went to many cocktail parties in their time, but they would have had a similar ability to detect and pick out their names against background noise. It would have been important in case they were needed to help get rid of an attacker or an enemy, quite literally helping themselves to survive.

Someone's staring at me!

A similar talent is the ability to somehow 'know' if a person near by is staring at us, even if we can't see them. Have you ever felt yourself being watched, then looked up and realized that someone was, in fact, looking at you? Scientists have tested this effect on animals, which also seem to know when they are being watched. USP is obviously very important in the animal world, where the ability to 'feel' that another creature is watching you could easily save you from being their next meal!

Some communities have highly developed skills of USP. The Cuma Indians, who live in the San Blas Islands off the coast of Panama in Central America, are said to use odour as a way of helping to judge one another's mood. When they meet, they clasp each other under the armpits and then sniff their palms! Our version of this is to shake hands, but we too are extremely aware of smell. Subtle odours certainly influence our feelings towards others, even if we don't know it. The Cuma Indians have taken this a step further and have developed their sense of smell as another method of communication.

So much information is coming into our

brain all the time – images and sounds, signals and smells – that it has to choose which ones are important and act as a kind of filtering system. But if we sense something that is really important, our brain automatically focuses on it and filters out all the other stuff. If we could learn how to develop this ability even more, we might all be capable of much more 'mindpower'.

Sensing Others

It's not uncommon to have strong feelings, that you can't quite work out, about unknown situations, people and places. Sometimes these are bad feelings – and for some reason your mind tells you that you should be careful in a particular situation, even if you can't see an obvious reason why. This feeling is called 'intuition', and it is a kind of USP that many people trust in. It's also a type of survival mechanism – another way of protecting and looking after ourselves that has probably been with us since ancient times.

Evidence also exists that shows that some people are able to 'tune in' to the feelings of others; somehow they sense it if a person close to them is ill or in trouble. There are many reports of this happening with twins, some of whom seem to have an ability to sense each other's feelings. There have been many reports of twins, thousands of miles apart, but one 'knowing' if the other one has had an accident or has been rushed to hospital. If this ability is genuine, it may have something to do with the fact that the twins developed together in the womb. They may have grown some unknown form of communication before they were born which has remained with them for life.

In some ways these events may be similar to crisis apparitions (ghosts which appear when a person is in trouble). Again, this seems to have something to do with a strong emotional link between people who are friends or relatives. All these cases could be 'explained' away by saying they are simply cases of coincidence – but, until we know more about USP and its effects, we can't be certain.

'Telepathic' Animals

Some animals have incredible skills that seem quite spooky at first sight. Dogfish, for example, know if their favourite meal of flatfish is near by, even though the flatfish are completely invisible, hidden under the sand of the seabed. This is because the dogfish have highly developed senses that can pick up the tiny muscle-movements of their hidden prey. Some species of eel can also detect nearby creatures by means of a kind of invisible 'net' which surrounds them. This is a form of electromagnetic field that can detect the presence of other creatures near by, a little like radar.

Another interesting skill is the ability some animals have of 'predicting' earthquakes long before they happen. Scientists believe that certain species are very sensitive to vibration. Dogs, for example, can hear sounds beyond the hearing range of humans and, in a similar way, can pick up vibrations long before humans sense them. This is why there are many reports of animals running away in places where an earthquake happens – long before any human feels the first tremor.

Even though some of these effects seem very strange, even supernatural, there is really nothing spooky about USP. It's the result of these animals having developed in a certain way and having been provided with the tools they need to survive in their environments. And, as with most skills, some creatures have greater ability than others.

Testing Times

During the past hundred years or so, many, many experiments have been carried out in an effort to pin down real telepathy and the unknown 'power' of the mind – the sixth sense that is known scientifically as ESP. As with every kind of scientific experiment, ESP tests must be made under strict conditions to ensure that no mistakes are being made and that no cheating or trickery is involved. This isn't as easy as it sounds. Even if no one is deliberately cheating, there are many ways in which a test result can become flawed or inaccurate. One of the criticisms of many of the tests done in the past is that the conditions were not monitored thoroughly – so the results may not be absolutely correct.

These days, experiments have become more and more sophisticated, and elaborate machines can be used to try to obtain the most accurate results possible. However, there are some simple experiments which anyone can do to test their own ESP potential. These were the ones used in the early days of ESP research, and similar tests are still used by parapsychologists these days.

In the 1930s a researcher called Joseph Rhine first used what have become known as

‘Zener cards’ (named after their inventor, Karl Zener). There are five designs on the cards: a circle, a square, a star, a plus-sign and three wavy lines.

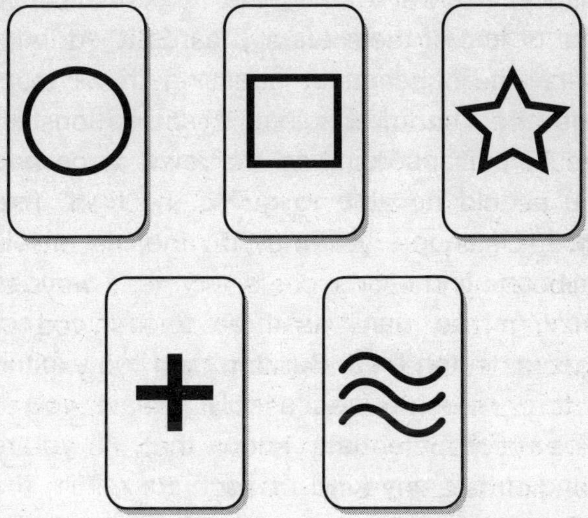

The five Zener cards

In this test the five cards are shuffled and put in a pile. A person takes the top card from the pile, looks at it and concentrates, trying to ‘send’ a picture of the card symbol to another person. The object is to see whether that person can identify the symbol on the card without seeing it. Try it – it’s easy to draw these images. Then persuade a friend to concentrate and do the test with you.

Obviously you might just be able to guess the symbol by pure luck – or what scientists prefer to call chance. But the element of chance is always taken into account in this kind of test. If there are five cards, then there is a 1-in-5 chance of guessing the symbol correctly. This means that, if you do the test five times, according to the laws of chance you should be able to guess the right card once. Of course you may do the test a few times and find your score is very high, perhaps you'll get as many as three or four correct guesses in ten tries. But don't get too excited – this doesn't necessarily mean you're telepathic! Scientists know that, if you're going to get any kind of accurate result, the test must be repeated over and over again, sometimes hundreds or thousands of times. It's easy to have a run of successes if you're only doing a few experiments, but to keep getting high scores with thousands of tests is much, much harder.

Prove it to yourself. Take a coin, decide whether you want 'heads' or 'tails', then toss the coin in the air. You might get it right a few times in a row; but if you tried tossing the coin 1,000 times and recording how many heads and how many tails you got, you'd probably find that heads would turn up about 500 times, and so would tails – proving that there is a 1-in-2 chance of the coin landing on either side.

In ESP experiments, the researchers work out what the average scores should be over a huge number of tests. If someone manages to get a very high score, it could mean any of the following:

- **It's a one-off, what is known as a 'fluke' (an amazing coincidence). Incredible, but these things happen sometimes.**

- **The experiment itself may not have been done correctly.**

- **It could be a case of genuine telepathy.**

Whichever conclusion the researchers come to, it's certainly true that in some trials people have obtained remarkably high scores. On one occasion a person being tested scored 588 successes or 'hits' from just over 1,800 trials. This is a success rate of 32 per cent. Although this may not seem anything to get excited about, it means that the person being tested got it right about every third card over hundreds of tests, which is pretty amazing.

Despite some people's incredible test results over the years, it is very difficult to prove that ESP definitely exists. People are always ready to criticize or find fault with the way the tests are performed. It is also very difficult for scientists to get people who may be telepathic to 'perform' again under scientific observation. All scientists have to be able to repeat their results in order to be taken seriously – and so far this has proved impossible with ESP.

How Could It Work?

If or when it is proved that telepathy *does* exist, how can it be explained in the world of science? One possibility concerns what are called wavelengths.

The range of colours that we all see, the radio and TV signals that allow us to tune in to our favourite programmes and the X-rays that produce pictures of your inside are all part of what is called the 'electromagnetic spectrum'. This is the name for all the invisible radiation that surrounds us and enables us to see and hear colours and sounds. The radiation consists of various-sized waves, a bit like ocean waves. At one end of the spectrum there are short wavelengths, and at the other end there are long wavelengths (such as the ones that carry your favourite radio station). Scientists believe they know the limits of the spectrum, but researchers into the paranormal suggest that other wavelengths may exist – 'telepathy' wavelengths. These could be one of two extremes, either much longer or much shorter than the wavelengths we know of. It is these wavelengths, they say, that carry the messages telepaths are able to detect. So far, however, no scientist has managed to discover them. This doesn't prove they do not exist – it could just be that our instruments aren't advanced enough to pick the waves up.

There is also a theory that the human brain could act like a phone-tapping device. This is a tiny machine which is attached to the telephone and is used by spies to listen in on people's phone conversations. But how could this work?

The human brain contains about 10 billion brain-cells, called 'neurones' or 'nerve cells'. That's more brain-cells in one head than there are people in the entire world! Each of these cells is connected to thousands of others, making the human brain the most complex machine we know of.

All neurones are linked by long tendrils that almost – but don't quite – touch at their ends. A signal sent by one neurone passes along the tendrils and crosses the tiny 'gap' to another neurone. This is how our thoughts and emotions are formed and transmitted through the vast network of our brain. All this happens in about a millisecond (one thousandth of a second) – that means the whole process can occur a thousand times in the blink of an eye!

Some researchers have suggested that, if telepathy is possible, it is because the signals crossing the gap between brain-cells somehow 'leak'. These stray signals are then picked up by telepathic people. Though the effect would be like a secret agent tapping a phone, there is one very big difference between phone-tapping and telepathy: the brain is far, far more complex than even the most elaborate telephone system.

Phone-tapping devices work either by intercepting part of a signal that is travelling along a single wire or by using a specially designed remote receiver to pick up a signal. In a similar way, a telepath would have to pick up a signal travelling between two neurones. The problem is that even a simple thought or instruction, such as deciding to blow your nose, requires many thousands of neurones working together. A single 'neurone tap' would gather almost nothing of any use – a telepath would need to 'tap' into many neurones all at once in order to get enough information to understand the whole message.

But how on earth could the telepath manage to 'decode' so many signals? How could the telepath manage to decipher all the trillions of impulses racing through another person's brain at any given moment? The telepath may want to find out what someone is thinking about him – and all he can pick up is a signal from the brain passing on instructions to scratch a leg!

It's clear that, although some of these ideas sound great in theory, investigators are a long way off proving them scientifically.

Moving Heaven and Earth

A favourite character in some science-fiction programmes is the one who can move things just by looking at them. Though no one we know has managed to do this in real life, some people say they can actually influence objects by the power of their minds. Such a talent is often referred to as PK (Psychokinesis) and throughout the past 100 years it too has been studied by many researchers of the paranormal.

The Luck of the Dice

The earliest serious attempt at studying PK was again down to Joseph Rhine in the 1930s. He got interested in the subject when a young gambler told him he could influence the way a pair of dice fell – by willpower alone. Rhine immediately started conducting thousands of tests on scores of people to see if this was possible. He asked the participants to concentrate hard and try to 'will' two dice to fall on numbers that would add up to a score of *more than* seven. It is slightly harder to get this score than a score of seven or *less*, as there are more combinations of dice which add up to lower scores (21 chances, as against 15 chances for eight or over).

As with his telepathy tests, Rhine found that most people got the results he would have expected, according to the laws of chance. Once in a while, though, a certain person would get a much better-than-average score. On a few occasions, he found people whose scores were so much higher than the average that the chances of this happening by accident were millions-to-one against!

In one set of experiments Rhine conducted 6,744 tests. By the laws of chance, these should have produced 2,810 successes – but one person got an amazing score of 3,110. The chances of achieving this are one in a billion!

But did Rhine manage to prove anything by all this?

Unfortunately for him, some of his methods were criticized by other scientists, and many still believe that Rhine did not conduct his experiments according to proper scientific methods. Others believe that his incredible results were just flukes – which do happen sometimes. Rhine himself argued that his tests were strictly monitored and that there was no trickery or influence involved. All these years later, it is still impossible to find out whether or not this was true.

What we do know is that, after nearly 30 years of experimenting, Rhine himself became convinced that some form of telepathy *does* exist. He decided that there is some kind of force in the human mind which can affect physical matter directly – but he could not explain how it worked.

Since these early tests, many experiments have been conducted by teams of researchers around the world – but there is still no absolute proof that people can influence objects by the power of the mind alone.

Little by Little

One of the difficulties for scientists who are trying to find out whether PK exists is that the effect is so tiny it is almost impossible to measure. It is because of this that researchers have named the effects they see in their laboratories 'micro-PK'.

In the 1970s a psychologist called Gene Glass came up with a revolutionary approach to studying results from PK tests. Because the effects of PK were so small, he realized that if he wanted to prove anything he would need to look at a very large number of results from different experiments. In fact, the smaller the effect, the more results he would need. It is a bit like tuning a radio. If a signal is strong, such as picking up your local radio station, it is easy to find. If, on the other hand, you are trying to pick up a weak signal, you may have to spend some time 'fine-tuning' the radio before you can detect it. This 'fine-tuning' is equivalent to conducting a large number of 'samples', or tests.

So Glass's idea was to get together *all* the data from *all* the tests that had been conducted over the past years. It was a difficult task because the tests went way back to the 1930s – and every one was different.

But eventually he found a way to combine them all. This allowed him to look at tens of millions of results drawn from experiments performed over a period of over 60 years – and to try finally to answer the question that had been dividing scientists during all this time. He came up with some interesting facts about PK:

- **Some (rare) people seem to have the ability to perform PK. They may be genuinely telepathic.**

- **PK appears to work best when the person is not deliberately trying to make it happen, or when they are concentrating on something completely different.**

- **In some small way, we may all have the ability to perform PK.**

For most of us, the last of these facts is the most interesting – we'd all like to have some sort of telepathic power, perhaps to help us win the lottery or do brilliantly well at exams. In reality, however, PK may have a much smaller effect over our lives than this. For example, have you ever experienced times when things have gone either fantastically well or terribly wrong? We all have good days and bad days – and many of us have experienced what's known as 'beginner's luck'. These are moments when you can do no wrong. It is on these occasions, the believers say, that micro-PK is working at its best, and these are exactly the times when, as Glass pointed out, we aren't really trying to make it work.

The Great Escape

A psychologist called Rex Stanford conducted a different type of experiment to illustrate the effects of micro-PK. He placed his volunteers in a locked room and asked them to complete a series of boring tasks. Without telling them that he was doing so, in another room he installed a machine called a 'random number generator'. This can churn out sequences of completely random numbers, just as computers can choose a sequence of lottery numbers for you. No one can possibly know what these numbers are going to be.

The volunteers could stop their boring tasks and leave their room only when the generator next door produced a particular sequence of numbers. If your family have ever done the lottery, you'll know that there isn't much chance of your numbers coming up – and it was exactly the same with this experiment. Professor Stanford calculated that they would probably have to wait several days for the machine to produce the right sequence. But, to his amazement, he found that some people managed to get out of the room within 45 minutes! How did the machine produce the correct numbers so quickly?

It seemed that it was the effect of micro-PK at work. If the people locked in the room were somehow able to influence the way the number generator worked, they could get away from their boring jobs and escape the room. And, as they didn't know about the number generator, they couldn't have deliberately been trying to 'will' it to choose the right numbers. As Glass had described, the effects may work better when you're not even trying.

Uri Geller – Spoon-bender

One of the best-known cases in which science has been challenged by someone who seems to have genuine powers is the example of Uri Geller, spoon-bender.

Uri Geller is a telepath who claims he can perform PK on everyday objects. His speciality is to take a spoon and bend it using purely the power of his mind. He simply rubs it gently between fingers and thumb, and the spoon bends. As well as doing this, he says he has other influences on objects – such as the ability to make watches and clocks stop and start.

Uri Geller first appeared on TV in the early 1970s, when he became an overnight sensation. He bent an assortment of spoons and metal objects in front of the cameras, causing people throughout Britain to rush for the kitchen drawer to see if they too could do it. Many of them said that metal objects in their homes had bent and watches had stopped while they were watching Uri on TV!

Since then, Uri Geller's skill has provoked strong feelings. Some people believe he is a talented telepath, others think he is just a talented trickster. To his credit, he has often allowed himself to be studied closely by scientists and, so far, no one has been able to explain how he does what he does – or how he might be tricking everyone.

Uri Geller has travelled around the world and has undergone tests at laboratories in Japan, France, Britain and the United States. He has been video-taped making objects appear from thin air, stopping machinery – and, of course, bending and distorting a vast assortment of cutlery, not only on film but before large audiences on live television shows.

A team in France have used a very powerful microscope, called an 'electron microscope' (which can magnify things hundreds of thousands of times), to study the spoons before and after Uri Geller has bent them. Testing the structure, weight and dimensions of the metal has shown that there appears to be a hardening around the area where the bend occurs. This is thought to be similar to the effect of applying intense pressure from outside. But as the only pressure comes from Uri's finger on the spoon, this cannot be explained. Other tests have shown that, as the bending occurs, the metal around the bend takes on a consistency similar to chewing-gum. Scientists are left baffled by the results.

Uri Geller definitely makes spoons bend, but to this day no one knows how he does it. He gives the impression that he himself does not fully understand his own abilities, and he appears to be totally convinced that what he is doing is paranormal. The world of science has not been able to prove any trickery – and he is certainly a walking phenomenon.

What's in the Future?

A small team of scientists in Scotland are, at this moment, working hard to prove that ESP and PK actually exist. They are all members of the Koestler Institute at Edinburgh University, named after the science-fiction writer, Arthur Koestler. So keen was he that scientists should find out more about the paranormal that, when he died in 1983, he left £1.6 million in his will to pay for the Institute.

The team perform experiments on volunteers to test their telepathic abilities. A typical test is called a *ganzfeld* experiment. In this test, the volunteer is locked into a sound-proof room with halves of ping-pong balls taped over his or her eyes so all that can be seen is a soft, glowing light. A soundtrack called 'white noise' is played in the background. This may sound a bit strange, but the idea is to put the person into a relaxed state that will prevent their thoughts from interfering with the act of telepathy. Then one of the scientists tries to 'beam' a particular image into the volunteer's mind (the image is chosen at random by a computer in order to avoid accusations of trickery). The results are then recorded.

In just one set of experiments on more than 100 people, the researchers found that volunteers had a success rate of 50 per cent – which is twice the 25 per cent rate expected. The chances of this happening by a lucky coincidence are 1 in 14 million!

The team believe that they now have the best evidence yet that ESP does exist. Despite all their findings, however, it will be a long time before they are taken seriously by the rest of the scientific world. But they will continue to work on ideas for new tests – and they are sure they are getting much nearer an answer that will satisfy everyone.

The Unexplained

Despite all the research done over the years, scientists themselves disagree strongly with one another over the whole issue of ESP and PK. Some accept that the phenomena do exist but that they have no idea how they work. Others say there is no such thing as ESP and any effects are caused by natural phenomena such as coincidence or errors in the experiments.

Clearly, something is going on that we don't fully understand. As yet, there is nothing in the known laws of physics that can explain how telepathy works – and, of course, scientists always want clear explanations for things. But should we dismiss things just because we can't explain them?

Until an organization like the Koestler

Institute comes up with a sound scientific explanation, it will have a hard time convincing everyone else. Until then, the power of thought will have to remain in the file marked 'Unexplained'.

Chapter 3

Mind and Body

At the window.
Look into my eyes. Now empty the safe and give me the money.
Warren leaves the bank with a full bag.
People are stopped in their tracks by his hypnotic power.
But outside the bank Special Agent Schwarz is waiting.
I have the suspect in sight.
OK. I'm ready to go.

Schwarz catches up with the suspect.
Warren Nobel, you're under arrest.
And Warren turns around to use his hypnotic powers on her.
Only to find that he has been foiled by agent Schwarz – who is wearing a specially designed pair of wraparound shades that protect her from Warren's mesmerizing stare.
That trick won't work with me, Nobel. Now get in the car!

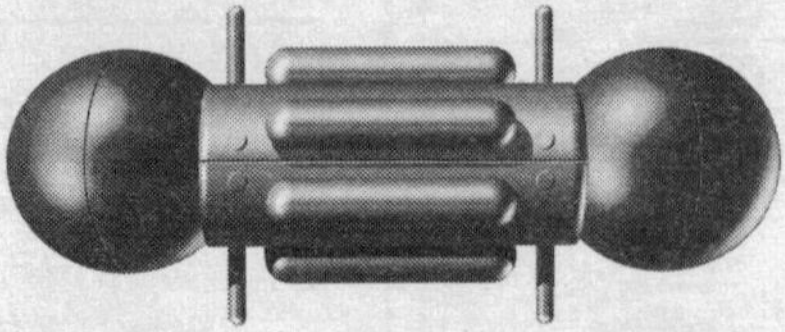

Mind and Body

THE FACTS

Your brain is an amazing organ. It controls your body, telling the different parts what to do every second of the day. Even a simple action like making a phone call requires a whole series of commands from your brain. Not only do you use it to keep everything in your body working, to do your maths homework and to read books like this, it also has some remarkable hidden secrets.

In science fiction you may have seen people who seem to have amazing abilities, people who can control other humans and make them do whatever they want. But in real life too, some people have been able to use and 'train' their brains and bodies to do some extraordinary things.

Secrets of Hypnosis

'Look into my eyes . . . You're feeling very, very sleepy . . . You will now do anything I tell you to do . . .'

Hypnotism does seem pretty spooky – especially when sci-fi characters stare into a person's eyes and seem to 'brainwash' him or her. You may even have seen hypnotists on TV or live on stage and wondered how on earth they do it. A member of the audience is put into a kind of 'sleep', called a trance, in which it seems as if their thoughts and actions can be controlled by the hypnotist. Is this some kind of trickery – or do some people really have an amazing power over other people's minds and bodies?

Hypnosis is actually a very old technique. It was developed in the eighteenth century by an Austrian physician named Franz Anton Mesmer; he treated many of his patients using this method. He became incredibly famous during his time for using hypnotism – in fact, hypnosis used to be called 'mesmerism', after him.

Though it looks mysterious, hypnosis is not a very complicated technique. The idea is that the hypnotist relaxes someone, the 'subject', and then suggests things for them to do. This is how hypnotists in entertainment can make volunteers from the audience do all kinds of things, some of which can be very silly. Reports from such shows say that volunteers have done bizarre things, such as crunching away at raw onions, believing they're delicious apples. Sometimes subjects are hypnotized into believing that they are small children again – which looks hilarious to the audience but can be very embarrassing for the person after the event!

So how exactly does hypnotism work? A special part of our brain controls how we react to others and to ideas put to us by other people; it controls how willing we are to do the things other people tell us to do. The hypnotist's job is to get the subject relaxed; this helps stop this part of the brain doing its usual job so well. We all have shyness and fears (often called inhibitions), but the skilled hypnotist can dull the signals coming from the part of the brain which handles these feelings. The hypnotized person behaves in a more free and open way and is more likely to do things that the hypnotist suggests. This doesn't mean that a hypnotist can make anyone do anything he or she doesn't want to do – or anything that the person would find wrong or shameful in their normal state. The sci-fi idea of being 'brainwashed' is very unlikely. Most hypnotized people know what's going on – and not everyone can be hypnotized. Some people just aren't open to it.

Hypnosis *can* be great fun when used as entertainment, but it is not just a neat party trick. It can also be used to help people, both physically and mentally. Some doctors – and even dentists – use hypnosis to relax their patients, and this can be very useful if someone is nervous about undergoing treatment. Some people try hypnosis to help them strengthen their willpower if they want to do (or stop doing) something, like giving up smoking. Hypnosis has other practical uses as well.

Fighting Crime

There have been a few cases in which hypnotists have been asked by the police to help solve serious crimes. If you witness a crime, it can be very difficult to remember every detail of what you saw and heard. So some witnesses have allowed themselves to be hypnotized in order to help them remember things more clearly. The hypnotist asks the witness to relax and then suggests they 'go back' to the scene of the crime and re-live the experience. 'What do you see?' asks the hypnotist. And sometimes the witness can recapture every detail of the event, even if it had happened days or weeks earlier. In these cases, hypnotism allows the brain to focus better on that past memory – because it is not being crowded out by all the other thoughts that race around our heads during our normal daily lives.

Reliving the Past

Using hypnosis to take someone back to the past can produce all kinds of surprising stories. Investigators of the paranormal have asked people who believe that they have been abducted by aliens, to be hypnotized so that they can re-live their experiences. Sometimes these people have given amazing descriptions of the inside of alien spaceships and have described conversations they've had with aliens!

Hypnotism has also been used in cases of what is called 'reincarnation'. There are many people who believe in the idea that we live, not just once, but many times on earth, and that our 'souls' move from one lifetime to another in new bodies. This belief is central to many religions, particularly Hinduism and Buddhism.

Some of the scenes described in these cases by people under the influence of hypnotism have been quite remarkable. A few people have managed to recall dozens of different lives before their present one and have given detailed accounts of their experiences, some of which took place hundreds of years ago.

One famous case involves an Indian boy called Ravi Shankar who described the toys he had played with when he lived, years before, as a boy called Munna. Munna's family had a house in the same city. When investigators tracked down the family, they found the toys to be exactly as Ravi had described them. Even more amazing was the fact that Munna had been murdered – and Ravi was able to describe exactly what had happened without knowing any of the detailed information held by the police!

Although at first these accounts seem to prove that alien abduction and reincarnation were the true experiences of these people, good scientists have to look very closely at the evidence. In both cases, the hypnosis may not be what it seems. First, anyone describing a past life could simply be remembering something experienced as a child and since forgotten. It is only under hypnosis that these memories return – perhaps stories read when aged five or six, or a programme seen on television when even younger. If these long-forgotten images are recalled, it is easy for the person to believe that he or she had these experiences in person.

Another drawback is that ideas can be planted in our heads without us knowing it. When we are hypnotized we become more open, and more easily accept, ideas put to us. The investigator can accidentally (or, in some cases, intentionally) plant ideas in our minds. These suggestions can then grow and become real to us, so real that we become convinced that something imaginary really did happen to us. This is called 'false memory syndrome'. It has caused many doctors to wonder whether what people describe in such investigations really is the truth.

Mind over Matter?

The relationship between our brain and our body is a complex one. In some ways, our brain has a great deal of control over our bodies. What you tell your body to do, it does. But there are also situations when this doesn't happen. No matter what your brain may be saying to your body, you can't control what your body does or how it reacts – for example, it's hard to stop yourself blushing or prevent your hands from shaking when you're very nervous. And there are some things, like the shooting pain you get when you stub your toe, which no one has any control over – or do they?

Some people do believe that they can control their bodies to an amazing extent – and that they can even control the amount of pain they feel. Such people need to train and practise secret arts for many years in order to be able to do the things they do. But, once they have completed the training, they can perform some truly astonishing feats.

These incredible skills have been practised since ancient times, and they probably originated in the East thousands of years ago, in ancient India and Tibet. The ability to control the body is an important part of some religions, and those who practise these ancient arts are seen as great holy men with very special powers – called yogis or fakirs.

Amazing Body Feats

There are strange tales which tell of yogis and fakirs sitting in one spot in the 'lotus position' – a special posture in which the legs are tucked up under the body – not for weeks or even months, but for years! They do not move, they do not eat or drink or speak. Other stories tell how these holy men can sit under water for hours, or they balance on a tall, narrow pole without moving a muscle – also for years at a time! Many of these tales cannot be proved – they are handed down from generation to generation and are probably greatly exaggerated along the way. But there certainly are people alive today who can do the most remarkable things with their bodies.

You may have heard stories of people being able to lie down on top of beds of nails without hurting themselves. Some yogis can actually do this – and they don't seem to feel pain. Witnesses say that when they get up from the bed their backs are unmarked by the sharp, needle-like nails. How on earth can they manage this?

The way they do it is through a mixture of training, skill and a few good tricks. Lying on a bed of nails requires great concentration. This can be acquired only through deep meditation and many agonizing years of practice. The yogi has to be totally relaxed so that his muscles do not tense and allow the tips of the nails to penetrate the skin. The yogi also has to get on and off the bed in a very special way – the secret of which has been passed down by earlier generations.

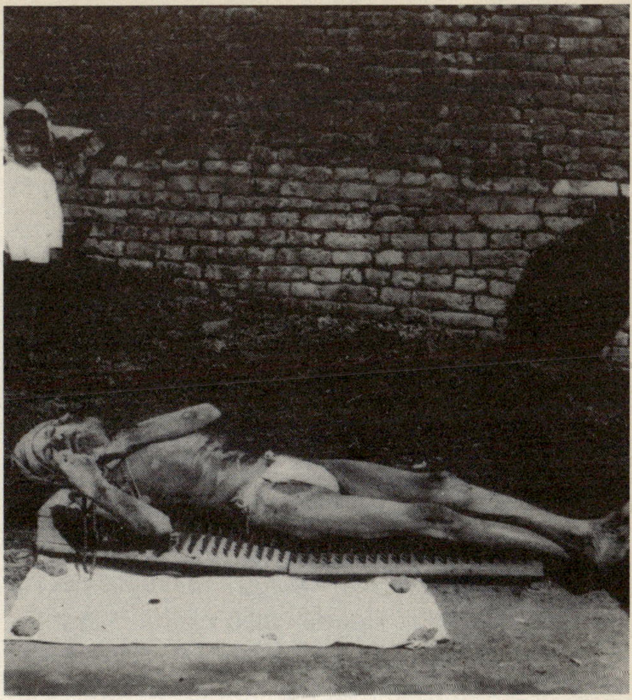

A holy man lying on a bed of nails in Calcutta, India.

The Great Houdini

Although most fakirs and yogis live and perform in India and the Far East, there have been some examples of westerners who have managed to acquire some of their skills. Probably the most famous of these was a man called Harry Houdini, who toured the music-halls of the world, enthralling audiences with his sensational act, from the beginning of the twentieth century until his death in 1926.

An escapologist is a person who entertains people by escaping from extremely difficult situations – and this is what Houdini did for a living, regularly escaping from strait-jackets in seconds. He also survived being immersed in freezing rivers with his hands and feet bound by manacles – claiming that he had defied death this way over 2,000 times during his career! On one occasion he managed to survive in a sealed coffin that was submerged in a swimming pool for an hour and a half, beating the record set by an Egyptian fakir. Houdini also claimed that he could take any blow anywhere on his body as long as he had time to prepare for it. Ironically, he died from a ruptured appendix – after a fan punched him in the stomach before he had time to tighten his muscles.

Walking on Fire

One of the most spectacular feats of the yogi is the ability to walk on fire. This has been practised in the East for thousands of years, but it was first witnessed by westerners in the twentieth century, when tales of amazing fire-walks were brought back by adventurers who had travelled to the East in the early 1900s. Those who 'fire-walk' usually walk short distances over extremely hot coals.

In many cultures, the courage and control needed for such acts as fire-walking are treated as a religious ritual, a kind of cleansing process. In ancient times, facing such terrors was also used as a form of judgement – the traditional 'trial by fire'. According to one story, in Italy in 1062 the Bishop of Florence was accused of corruption by the saintly Peter Aldobrandini. Peter issued a challenge: that they should both walk on fire; he believed this would give the Bishop an opportunity to prove his innocence. A corridor of hot coals was prepared and a bonfire placed at each end. Peter Aldobrandini entered the flames, walked the path of scorching coals and came through the fire at the far end totally untouched. When it came to the Bishop's turn, he (very wisely!) turned down the opportunity and decided to resign instead.

Fire-walking has been demonstrated many, many times, and there is no doubt that *some* people can perform this incredible feat. But again, although it is an extremely dangerous thing to do, the secret of fire-walking has a lot to do with talent, training and tricks.

Secrets of fire-walking

The walker has to have hardened skin on his feet. Yogis never wear shoes, and after many years of walking in bare feet they develop a thick protective layer of hard skin on the soles of their feet.

Although very hot, the runway in a typical fire-walking session isn't usually more than about 350° Celsius. Of course, this is hot enough to burn away flesh, but the ashes and cinders on the path aren't very efficient at conducting heat. What this means is that only a little of the heat contained in the coals is transferred to the feet of the walker.

The walkers don't really walk – quite understandably, they run! This means that the heat has very little time to get through to their feet. A typical fire-walk is about 3–4 metres or yards long and lasts a very few seconds.

The final protection comes from something called the 'Leidenfrost effect', named after the man who discovered it. Professor Leiden discovered that liquid exposed to a sudden, intense burst of heat

creates a layer of steam above it. What this means for the intrepid fire-walker is that the heat from the coals evaporates some of the water on the soles of his feet. This then creates a protective layer of steam between him and the fire.

Of all these facts, the most important one is that coals are not very good at conducting heat. Different materials can be at the same temperature, but they won't all feel as hot as each other. If you picture yourself trying to walk across a beach on a hot summer's day, you'll see this effect. As you begin to walk, you feel the sand burning your feet, and you dash for a beach mat. This is cooler because it doesn't conduct heat as well as the sand does. Then you run on to the pavement and, although the sand feels hot again, the concrete burns like mad!

The pavement is so hot because concrete is a better conductor of heat than sand or straw. For this reason, you never see fire-walkers running across concrete heated to 350° Celsius or, worse still, a metal sheet heated from below! But this still doesn't fully explain how most people would burn their feet if they tried to fire-walk, while skilled fire-walkers show no signs of burning. Research has shown that this amazing resistance to heat and pain may also have something to do with the brain's control over the body – a real case of mind over matter.

Having a Brain Wave?

As already explained, your brain is composed of billions of special cells called neurones which connect with one another to produce all our thoughts and actions. These neurones communicate by electrical impulses, and some of this energy 'leaks out'. These leaks are known as 'brain waves', and they were first discovered in the nineteenth century.

There are four distinct types of rhythm produced by the brain, and each one relates to a different mood or 'brain state'. So, if you are relaxed your brain produces one type of wave, but if you are very busy doing your homework, you produce a very different type. These brain waves can be detected by a machine called an EEG (short for 'electroencephalograph'), which picks up the tiny electrical impulses between neurones in our brains. The patterns that the impulses create can then be printed on a graph. Here's what the four types of brain wave look like.

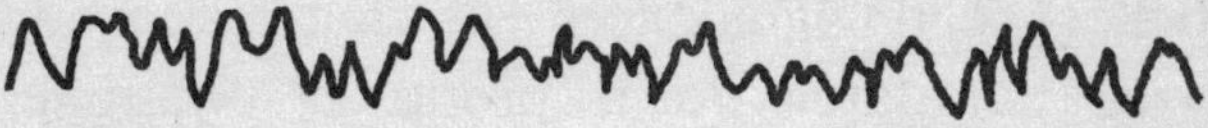

Alpha-waves, produced when we are resting or relaxed.

Beta-waves, produced when we are active, especially when the brain is working, solving problems or doing something physical.

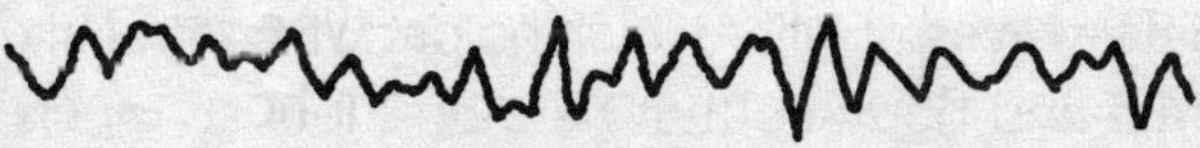

Delta-waves, produced during sleep.

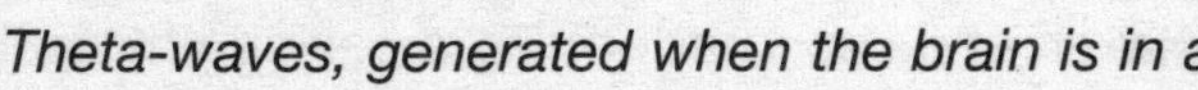

Theta-waves, generated when the brain is in a deep sleep or in a trance state.

In the late 1970s a series of experiments was carried out on an Indian yogi who was wired up to a collection of monitors. His brain responses were compared to a group of 'normal', untrained volunteers who were being subjected to the same treatment.

The experimenters applied electric shocks to the fakir and to the volunteers and monitored their heart-rate and brain-wave patterns at the same time. They found that, after a short time, the volunteers could take no more and looked exhausted, whereas the fakir remained calm and controlled. They also discovered that his heart rate was much slower and that he had produced strong theta-brain-waves throughout the experiment. The scientists came to the conclusion that the yogi was able to produce theta-waves at will – which helped to control his brain and body so that he didn't feel so much pain. This meant he was able to hypnotize himself into a relaxed state. This ability has also been noted by people in other situations.

Human 'Super-powers'

Stories of people slipping into a trance-like relaxed state and performing amazing physical feats at times of high stress are surprisingly common. This can sometimes happen in a dire emergency – when we may be called upon to carry out tasks we would never be able to perform under normal circumstances. Many tales are told of people freeing loved ones from cars by bending metal supports with their bare hands or rolling cars over in order to free those trapped underneath!

One woman drove her car over a cliff, and then managed to crawl out of the wreckage that had squashed her into a space only a metre (a yard) square. Despite having a broken arm, severe concussion and facial injuries, she managed to walk a kilometre (that's more than half a mile) to the nearest telephone, where she calmly called her father and waited for an ambulance to arrive. Today, she has almost no memory of the time that elapsed between escaping from the wreckage and reaching hospital.

How does the body make itself do this? At times of stress, a chemical called adrenalin is produced in the bloodstream. This helps slow down the flow of blood if you are bleeding. It also provides emergency supplies of energy to your body, which you may need to help you get out of a tricky situation. All of us will produce adrenalin if necessary. But this doesn't mean we could all develop 'super powers' in an emergency. Some people seem to have the ability to influence their brainwave patterns, particularly the relaxing theta-waves, just as yogis do.

Animals too can sometimes fall into this 'self-hypnotized' state. This can happen when an animal is cornered and trapped by a predator. Instead of trying to fight back or to struggle, the terrified prey sometimes drifts into a trance. Like the Indian yogi, its brain will produce theta-waves so that it will suffer less pain when it is killed.

Self-hypnosis and Meditation

Many people believe that the trance-like state described above, which is a form of self-hypnosis, is something that can be learnt. The technique requires concentrating on a simple fixed image (a candle is a favourite), which then allows the person to focus the mind and channel the brain-wave patterns into a slower rhythm. These are usually alpha-waves (normally associated with sleep) or theta-waves.

One form of self-hypnosis that people use is called 'transcendental meditation'. This became popular in the West during the late 1960s, especially when the most famous pop group of the time, the Beatles became interested in it. TM, as it is called, is used by many people as a way to relax. For the real enthusiasts, learning the art of TM is a way in to the more advanced skills displayed by yogis and other holy men.

The most important aspect of TM is your 'personal mantra'. A mantra is usually a simple word that is chanted silently over and over again – a word you must never tell anyone else, otherwise it will lose its power and the technique will be destroyed. This repetitive chanting is a good method for putting your brain into a relaxed state by reinforcing a single, simple rhythm.

Mind and Body Together

Although feats like being able to walk on hot coals require years of training and demand special natural talents, to some extent we are all capable of displaying amazing mental and physical powers. Just think how incredibly difficult it is for someone to play an entire piano concerto from memory, or the great stamina that is needed to run a marathon. The brain is the most complex machine we know of, and when it works in perfect harmony with the body the two can work wonders.

Scientists and doctors are still discovering new facts about the brain, and they believe that there is a great deal more to find out about the relationship between our mind and our body. They are looking at areas such as disease, trying to find out how close a link there is between our health and our state of mind. In the past the brain and the body were treated as two completely separate things, but these days many people believe that the two are more strongly connected.

There is no doubt that the mind can allow some of us to influence our bodies in amazing ways. And we must wait and see whether scientists can find out more about harnessing these incredible abilities.

IMPORTANT: Do not try to imitate any of the feats described in this section in any way. They can be extremely dangerous.

Chapter 4

Playing with Time

The line goes dead.
She doesn't believe me. What am I going to do?
There's only one thing he *can* do.
Flagging down a taxi, John rushes to the airport.
Oh no – there's flight CTX09 on the screen. STOP THE PLANE, IT'S GOING TO CRASH!
Arrest that man. He's causing a panic!
John is taken to the police cells.
Please believe me! I'm not crazy.
Meanwhile . . .
We have an emergency situation . . . repeat . . . emergency . . .

The disaster happens just as John said it would . . . luckily the President makes it out alive .
PRESIDENT SURVIVES PLANE CRASH DISASTER
Meanwhile, in the airport cell, there's a knock at the door.
Who is it?
Special Agents Schwarz and Kelly. We'd like to talk to you about your dream . . .

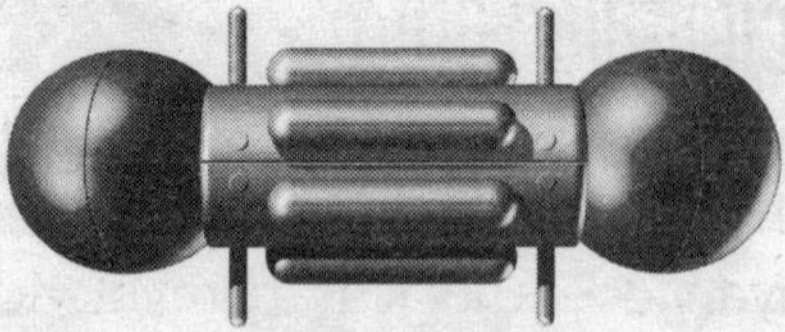

Playing with Time

THE FACTS

We usually think of time in three ways: as the past, the present and the future. The future is the unknown, which is why it is so fascinating – anything could happen to us tomorrow, next month or next year. People have always been fascinated by the idea of playing with time, whether that means having visions of the future, travelling back in time or changing their destiny in some way – all stories that are common in science fiction. But in real life some people claim they can really see into the future and predict what is going to happen. Others believe that it's even possible to 'slip' into different times and back again to the present. But what do scientists think about all this?

The Fascinating Future

The ability to see into the future and tell others what will happen is called prophecy, prediction or clairvoyance – and it is a mysterious and thrilling idea.

Clairvoyance has a very long and rich history, and it crops up in all sorts of places. In ancient Egypt the priests of the Sun-God, Ra, were prophets who could see into the future. The ancient Greeks consulted oracles, a place or a person where prophecies were given, to find out the best time to wage war or make peace. Throughout history, those in power have consulted 'soothsayers' (the old-fashioned name for fortune-tellers) for advice. In William Shakespeare's famous play, a soothsayer warns Julius Caesar of his impending murder with the famous line: 'Beware the ides of March' (the 15th of March).

The future has always been a popular theme in science-fiction films and TV series, such as *Back to the Future*, *The X-Files* and *Red Dwarf*. These often have stories about people trying to change the course of the future in some way. In one episode of *Red Dwarf*, the characters went back in time to warn their earlier selves of a future disaster! All kinds of weird and wonderful things can happen when characters in sci-fi have the ability to see and control their own futures.

The idea that we might be able to find out what could happen to us tomorrow or next year is certainly very exciting. It can also be very frightening, because nobody wants to hear bad news from a fortune-teller. However, one of the most famous clairvoyants in history was a man who predicted all kinds of terrifying world events, including war and famine, death and destruction.

The Predictions of Nostradamus

Nostradamus was born in France in 1503, and for many years he was a doctor. When his family died of the plague in 1534, his medical practice collapsed and he started his travels around Europe, gaining a reputation as a mystic and a healer. During this time he began to write verses which contained his prophecies concerning the future. He had them bound together in collections, completing almost 1,000 verses before his death in 1566.

To some people, Nostradamus's works are truly amazing. They firmly believe that he predicted some of the major events that happened in the centuries following his death – including both World Wars, the rise and fall of the British Empire, the coming of technology and even the end of the world itself!

Here is one of his verses. What do you think it could be predicting?:

They will think they have seen
the Sun at night
When they will see the pig half-man:
Noise, song, battle, fighting
in the sky perceived,
And one will hear brute beasts talking.

Followers of Nostradamus believe that this verse predicts atomic weapons and fighter aircraft in the twentieth century. Seeing the 'sun at night' could be a very good description of an atomic bomb exploding. And what of: 'When they will see the pig half-man:/Noise, song, battle, fighting in the sky perceived'? It is easy to picture a wartime pilot wearing an oxygen mask looking like a 'pig half-man'. But is this what *you* thought it meant?

As you can see, the problem with this verse and other predictions by Nostradamus is that they can mean different things to different people. His verses don't clearly describe exact images or events. They could have lots of different meanings. It's also easy for people to assume that Nostradamus was referring to an event such as the Second World War *after* they had lived through it. What would people living in the sixteenth century have made of these words? And if the work of Nostradamus had not been discovered until the twenty-second century, what would people living in the future make of them?

Even though Nostradamus's words are so unclear, it's certainly true that many famous figures in history have studied his verses and have been guided by his predictions. We'll just have to wait and see if any more of his predictions come true for the twenty-first century!

Astrology Truths

Probably the most popular kind of 'fortune-telling' these days is astrology. Many of us enjoy reading our personal horoscopes and finding out about our personalities in newspapers and magazines. These predictions and descriptions are written by astrologers, people who believe that our lives are shaped and guided by the positions of the planets in the sky.

Astrology is many thousands of years old and probably arose out of our ancestors' obsession with the stars in the sky. The stars have never lost their power to generate wonder in all of us. In ancient times people believed that the heavens were the home of the gods; today we look in awe at the stars because they are still beyond our reach – but also because they may be home to other life-forms. It's easy to understand why ancient mystics believed that the stars affect our lives directly. They created the system of astrology, which gives everyone a 'star sign' depending upon the month he or she is born. There are twelve star signs, or 'signs of the Zodiac', which are the constellations the Sun appears to pass through at different times of the year.

Followers of astrology believe that your star sign affects your personality and determines the kind of person you are – for example, whether you are shy or bossy, hard-working or lazy, and so on. Astrologers also believe that they can predict our future by using the changing positions of the planets as their guide. They can draw up a personal 'astrological chart' for an individual; this can be very detailed and takes note of the exact positions of the Sun, the Moon and the planets at the time and place of a person's birth.

Like the predictions of Nostradamus, it's easy to 'read' anything you like into most horoscopes. If you try reading several different horoscope predictions in a newspaper in one evening, you'll probably find that several of them, not just the one written for your own star sign, could just as well have described your day. Astrologers know that there are lots of things that all human beings have in common – for example, we all worry about money and we wonder about love. So there's a good chance that, on any one day, one or both of these things will cross a person's mind. As you've probably realized already, many horoscopes around today are made up by using pure guesswork.

But there are astrologers around who take their art very seriously, and they really believe that our destinies are controlled by the heavens. Is there any scientific basis in this?

Heavenly Bodies?

Astrologers have tried to explain how the procedure could work scientifically. The planets are held in their positions in the solar system by gravitational forces. Gravity is the invisible force that gives everything weight and keeps us 'stuck' to the ground. The gravitational 'push and pull' between the Earth and the Moon has a huge effect on the Earth's water This force creates the tides of the oceans which is why the times of high and low tides always depend on the position of the Moon. Because the position of the planets has such a major effect on water, astrologers argue that their position could have a similar effect on the human body – which is ninety per cent water.

Although these facts about the tides are true, most scientists have great difficulty in believing that the Moon can influence the human body in any way. There is no evidence that the stars – or even the planets close by in our solar system – can affect the water in our bodies or influence us as individuals. And even if they could, how would this enable us to make predictions about our futures? This idea has no basis in real science at all, and so far scientists are unconvinced that there is any truth in astrology.

Test the Clairvoyant

Apart from the daily horoscope, at the end of every year many newspapers employ clairvoyants to make predictions for the next twelve months. Most people forget about these predictions soon after reading them, but, if you check back and compare the predictions to the actual events of the following year, you'll notice that the clairvoyants aren't usually very accurate.

Try it yourself. In the library you'll find all the old newspapers from previous years. Try matching up the predictions, usually published in the last issue of the year, with the almanacs (yearly books) of events for the following year.

As an example, the following predictions were made by clairvoyants at the end of 1981 for 1982:

- **'The Cuban president, Fidel Castro is overthrown in a major uprising.'**
- **'The latter part of 1982 will see petrol prices rising.'**
- **'A controversy will break out when it is discovered that a man-made device sank the *Titanic*.'**

In fact, Fidel Castro is still in power, petrol prices fell in the second part of 1982, and no such discovery has ever been made about the *Titanic*.

But all the clairvoyants failed to predict these big events:

- **The death of the famous former actress and princess, Grace Kelly.**

- **The stock market hitting an all-time high in November 1982.**
- **The death of the premier of the Soviet Union, President Brezhnev.**
- **The first operation in which an artificial heart was placed in a human being.**
- **An attempt to explode a terrorist bomb in Washington, D.C.**

What all this shows is that much of what clairvoyants foretell is guesswork – and poor guesswork at that! This isn't something that you find just in newspapers. Sometimes, very skilful guesswork is used by 'fortune-tellers' to convince people that they should believe what the clairvoyants say.

Tricks of the Trade

There are plenty of so-called 'clairvoyants' around with tricks up their sleeves, so the next time Madame Zora peers into her crystal ball or reads your palm at the funfair, it's best not to take her seriously. Unfortunately, the whole idea of having your fortune told can really give some people the creeps – which is exactly the reaction the clairvoyants want, since it puts them in a position of power right from the start.

So how are these 'fortune-tellers' able to tell

you things about yourself if they are just fakes? Well, some people are very clever at reading signs – they are a bit like those people who have a form of USP (see *The Power of Thought*, page 34). They have trained themselves to notice and read signals that most of us are not aware of, like a detective picking up clues. Our body language and other subtle signals we don't even realize we're making can tell other people a lot about us.

These people are also very skilled at asking the right questions, leading us into answers that reveal a great deal about ourselves. For example, if they say something like: 'I see a recent loss in your life,' their tone of voice phrases it as a subtle question. If we have suffered a loss, we react, even if only in a tiny way, by a small flinch or a stiffening of the shoulders, which leads them to suspect that they have hit a raw nerve. Of course, the loss could be anything – lost money, a friend who has moved away, or the death of someone close to us. However, we get taken in and end up believing that the fortune-teller really knows all about us – when all they've really done is ask us lots of clever questions.

Predictions of Disaster

Prophecy and predictions are often linked to huge disasters, and many people declare that they have 'seen' terrible tragedies before they actually happened. The fact that predictions are often made before such catastrophes is not really surprising, because they are exciting and sensational – after all, who would take any notice if a clairvoyant predicted that Mrs Brown's cat would get stuck up a tree on Tuesday?

The 'Unsinkable' Titanic

One well-known case was the sinking of the famous ship, the *Titanic.* After she hit an iceberg and sank in 1912, several people claimed that they had decided not to go on the trip – because they already 'knew' the *Titanic* would sink. Of course, these stories could easily have been made up after the sinking took place. However, there was one incident where this couldn't have been the case. A man called William Klein warned his friend, before witnesses, not to travel on the ship. His friend ignored him – and died in the Atlantic when the *Titanic* sank. Another man, named W. T. Stead, was also given a warning by two clairvoyants.

He, too, ignored their advice and met an icy death.

Although these accounts may seem very impressive, we should look very carefully at the facts before making up our minds as to whether or not real clairvoyance lies at the heart of the mystery. There are two good reasons why predictions of disaster were made before the *Titanic* sailed.

1 Guesswork

Between 1890 and 1912, several novels were published about huge ocean-liners sinking in the Atlantic. One of these was Morgan Robertson's 1898 novel, *The Wreck of the Titan*. In this book, a ship called the *Titan* sank with great loss of life after hitting an iceberg in the North Atlantic in April. Fourteen years later, the real *Titanic* sank – after hitting an iceberg in the North Atlantic in April!

No one, least of all Mr Robertson, claimed he could 'see' the future, but the fact remains that he actually managed to predict the tragedy far more accurately than any of the clairvoyants who warned people about the *Titanic.* What this shows is that those who predicted the disaster were no more accurate than someone guessing – or, in this case, a writer working out the story for a novel. Morgan

Robertson based his ideas on the fact that the most popular time to sail to New York would be in the spring, allowing the passengers to spend the summer in America. This is also the time of year when icebergs are to be found far south and so are at their most dangerous. The name *Titan* was a coincidence – though not terribly surprising, as ships are often given very grand names like these.

2 The Titanic 'hype'

The other important fact is that the *Titanic* was a very unusual ship for her time. Her maiden voyage of 1912 was 'hyped' by the press as the cruise of the century and, because at the time she was the largest ship that had ever been built, everyone was talking about her. The *Titanic* was also claimed to be the 'unsinkable ship' – so it came as no surprise that a lot of people had their doubts about her!

In fact, we now know that that voyage of the *Titanic* was actually a very dangerous trip. The ship was the largest passenger vessel ever built and she was sailing at a dangerous time of year. The captain and crew were so confident about the trip that they may not have thought enough about safety procedures. The fact that there weren't enough lifeboats for all the passengers made the voyage even more risky.

Knowing the background to this event makes it easy to see how predictions of disaster got made – and how they were eventually proved right.

Real Predictions?

Despite our doubts concerning the *Titanic* stories, this doesn't mean that all predictions of disaster can be easily explained away. Predictions have been made when no one could have known – or even guessed – that a catastrophe was about to happen.

One example is the disaster that occurred at Aberfan in Wales in 1966, when a massive coal-slide destroyed a school, killing over a hundred people, many of them children. There were reports of people who believed they had had a vision foretelling this terrible event. Some of them even had witnesses – one woman told a friend about her vision about 45 minutes *before* the disaster took place. A psychiatrist who studied all these cases decided that some sixty of the stories sounded genuine. What can we make of events like these?

In some ways, these cases are rather like the 'crisis apparitions' described earlier, – 'ghosts' of real people who appear when they are in trouble of some kind. There are also similarities with cases of telepathy, when some people can 'sense' that a person close to them needs help or attention. All these cases seem to involve people picking up feelings and emotions from other human beings in trouble – but with different time-frames. Communicating by sending 'distress signals' from one person to another may be an ability that some of us have – but we are at a loss for an explanation as to how someone could pick up these signals *before* the event happens.

Travelling in Time

As you can see, traditional ways of foretelling the future, such as astrology and crystal-ball-gazing, don't really convince the world of science. But if there really *are* people who can see into the future, could there be other ways of looking forward which scientists might consider?

One possible way could involve the whole business of travelling through time. What if someone could get information about the future by actually visiting it? Another alternative would be if some sort of 'tunnel' linking different times could allow information to travel from the future back into our 'present'.

Scientists are now starting to think seriously about the theory of time travel. Not so long ago the very idea of travelling forwards or backwards in time was just a science-fiction fantasy. But today things are very different. Respected figures like Professor Stephen Hawking at Cambridge University are intensely interested in the whole idea of time travel and have written books on the subject. Now, this is not to say that we are on the verge of building a time machine – that would be brilliant – but, sadly, it is still a long way off. Scientists at present are just thinking about the theory – the mathematics of the possibility.

Time Tunnels?

One thing that scientists are considering is what is known as a 'wormhole'. Wormholes are an idea that science-fiction writers have always loved – but now they are a real scientific possibility.

Scientists visualize a wormhole as a kind of 'passageway' that links two random points in the universe, created when two black holes are joined up. They have been trying to work out whether wormholes could really exist and how they might be created. The whole idea is incredibly exciting because a real wormhole might act as a shortcut through the universe, enabling us to travel further through space than we'd ever dreamt of. (You can find out lots more about wormholes and space travel in the book *Sci-Fi Explained: Alien Life Forms*.) But wormholes may also offer an opportunity for us to obtain information about the future. Scientists are trying to work this out by thinking about how the two separate ends of the wormhole could exist in two different time-periods. If the hole linked the present time (ours) with the future time, could the events of the future be 'seen' at our end? If this were the case, the wormhole could act as a kind of time machine.

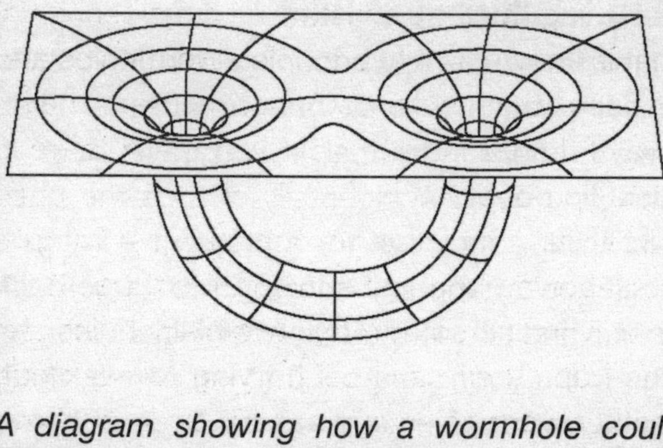

A diagram showing how a wormhole could 'link' two points in the universe.

Some scientists are even considering the idea that there may be lots of tiny wormholes that exist all around us. If information could travel through these wormholes, is this how predictions such as those made about the Aberfan disaster came about? Before you get too excited, you must face the fact that this idea brings with it lots of problems to be overcome – scientists don't yet know how we would find such a wormhole, let alone how we could 'see' into it. The whole idea may look good on paper, but we have yet to prove it by locating a real wormhole. However, as scientists discover more and more about this fascinating idea, who knows what they'll come up with one day?

Very Troublesome Time

If time travel ever did become possible, it could actually cause a great many problems. First, there is the danger that, if you travel back in time, you could accidentally change the past. Just think about that for a moment – imagine what would happen if someone could go back in time and kill the dictator Adolf Hitler when he was still a young man. If that happened, there would probably have been no Second World War, millions of people would not have died and the world today would be a completely different place – though no one knows in what way.

Another problem created by time travel that scientists have thought about is what is called the 'grandfather paradox'. This is similar to the story of the film *Back to the Future*.

The Grandfather Paradox

You travel back in time and meet your own grandfather when he was a young man. He is about to go on his first date with the girl who will one day be your grandmother. But something happens to stop him from going out that night (maybe your surprise arrival!). He then never meets up with your grandmother, they don't get married and you and your parents are never born. How, then, were you able to go back and talk to your grandfather when you don't exist?

Confusing, isn't it? One thing we can learn from this: if we ever do crack the formula of time travel, we will need to tread extremely carefully. As you can see, even a tiny change in one time can make a big, big difference later on!

Time-slips

Another situation in which some people say

they have found themselves travelling through time is through a 'time-slip'. The idea that ghosts may be images from the past that are 'imprinted' in the walls of a building and then played back was explored in Chapter 1. A time-slip is a similar idea. Some people claim to have suddenly walked into a different era, a kind of 'slip' in time – almost like appearing in a ghostly 'playback' scene. Amazing as it sounds, some of the stories are quite convincing. One famous case from 1901 described how two very respectable, middle-aged ladies claimed that they had 'slipped' into the past while walking in the gardens of King Louis XIV's palace at Versailles, near Paris. They even described how they encountered Marie Antoinette, the French queen of the 1780s!

More recently, during the 1970s, an entire family announced that they had stayed the night in what they described as an 'old-fashioned' hotel. All the other guests were dressed in 1930s-style clothes and used old-style currency. They were charged only £2 for the night's stay and even took photographs of each other outside the hotel. Unfortunately, the film they took came out blank, and when they tried to find the hotel again, a few days later, it was nowhere to be seen. Had they really slipped back into the world of the 1930s?

Others have reported seeing weird vehicles travelling very fast, both in the sky and along the ground. Some of these cases occur when driving along in a car. Suddenly a vehicle will shoot past at an incredible speed. Often human figures can be seen in the speeding vehicle, and sometimes these figures appear to be looking back at the witnesses in shock and amazement. Those who believe that time-slips are possible have suggested that these sightings are in fact advanced vehicles from our own future! The figures are also able to see us chugging along in our very 'old-fashioned' cars, so it seems that they too are experiencing a time-slip – into *their* past.

Fantasy or Fact?

So, what is to be made of all this? Some people say that clairvoyance, prophecy, time-slips and the rest cannot be explained by traditional science, mainly because there is no clear evidence to support them. But does this mean that we shouldn't take such things seriously?

The great science-fiction writer and innovator, Arthur C. Clarke, once said, 'Any sufficiently advanced technology seems like magic.' This is absolutely right. Just think for a moment what a scientist living in, say, 1850 would have thought if he had seen a machine that could produce a picture of the inside of a person. It would have seemed supernatural to him because science and technology in his time hadn't progressed that far. Only 45 years later, in 1895, X-rays were discovered, and soon after this the first X-ray machines had been built.

So, although for the moment science cannot explain many 'supernatural' phenomena, the search goes on. Scientists such as those at the Koestler Institute will spend their time probing the mysteries of the mind, trying to find rational, scientific explanations for them. The research on time travel will go on – and perhaps one day our technology will be advanced enough for us to build a time machine, something like the Tardis in *Dr Who* or the machine used in H. G. Wells's *The Time Machine*. Maybe human beings living in the future have already built one and are travelling back to our time and to other periods!

With such people working hard to discover what is really happening, one day they will find real answers to these great mysteries.

Glossary

Adrenalin A chemical that your body produces at times of stress.

ASSAP Stands for the Association for the Scientific Study of Anomalous Phenomena.

Astrology The (non-scientific) study of the positions of the planets, the Sun and the Moon, and their effect on humans.

Body language Communication without speaking, through movements and expressions of our faces and bodies.

Brainwave A movement of electrical energy in the brain.

Clairvoyance A power to see or understand things beyond the normal range of the senses.

Conscious Means alert and awake, also used for the part of your mind that is aware of your body and its environment
(*see **Subconscious***).

Crisis apparition A vision of a person in crisis or trouble which appears to another person.

Electromagnetic spectrum The complete

range of electromagnetic radiation, from the shortest gamma rays to the longest radio waves.

ESP Stands for Extrasensory Perception: communication of thoughts and feelings between people that are not understood by known scientific laws (*see* ***Telepathy***).

Hallucinate To see a vision of something that is not there.

Hypnotize To influence someone's mind into a state of relaxation and concentration, allowing deeper parts to be explored.

Intuition Getting knowledge about something, not in the usual way, but by instinct or 'feeling'.

Involuntary Describes a movement or function that your body performs without thinking and that it is not conscious of, e.g. breathing (*see* ***Voluntary***).

Meditate To think about something deeply particularly spiritual or religious matters.

Medium A person who claims they can communicate with the dead.

Neurone A type of cell that conducts nerve impulses that is found in the brain.

Paranormal Beyond normal explanation.

Parapsychology The study of mental

phenomena: things beyond the scope of normal physical explanation.

Poltergeist A noisy and badly behaved spirit.

Predict To foretell the future.

Prophesy To foretell the future through divine inspiration.

Psychic powers Powers of the mind which some claim to have but which are impossible to explain with the known laws of science.

Psychokinesis (PK) Influencing an object using mental powers.

Reincarnation The belief that when you die your soul is transported to another body.

Seance A meeting where people attempt to contact the spirits of the dead.

Senses Most people have five senses: sight, hearing, smell, taste and touch.

SPR Stands for the Society for Psychical Research.

Subconscious Describes the part of the mind that exists and functions without you being aware of it (*see* ***conscious***). Dreams may tell you surprising things about your subconscious feelings.

Supernatural Describes something that cannot be explained by natural laws.

Telepathy Communication of thoughts and feelings between people that are not understood by known scientific laws (see **ESP**).

USP Stands for Ultrasensory Perception: a form of mental power explained by science and possessed by some people who have great ability in using their senses.

Voluntary A function performed willingly or by choice (*see* ***Involuntary***).

Further Reading

You can read more about ESP and the paranormal in these books:

Andy Boot, *Elements of the Extraordinary: ESP*
Marc Gascoigne, *The Ghost Investigator's Handbook*
Jane Goldman, *The X-Files Book of the Unexplained, Volumes 1 & 2*
James Sasse, *And Now For My Next Life* (about reincarnation)
Colin Wilson, *The Unexplained: Ghosts and the Supernatural*

You might also enjoy these adult books:

John Fairley and Simon Welfare, *Arthur C. Clarke's Chronicles of the Strange and Mysterious*
John and Anne Spencer, *Encyclopedia of the World's Greatest Unsolved Mysteries*
Rosemary Guiley, *Encyclopedia of Ghosts and Spirits*
Michael White, *The Science of the X-files*

Index